I0087493

TAKE CHARGE

OF YOUR

MILITARY

TRANSITION

A 5 STEP ROADMAP TO
POST-MILITARY SUCCESS

ERICKA E. KELLY

https://erickakellyenterprises.com

MMXXIV Ericka Kelly, Ericka Kelly Enterprises, LLC

All Rights Reserved

Copyright © 2024 by Ericka Kelly. All rights reserved. No part of this publication may be reproduced, distributed, or transmitted in any form or by any means, including photocopying, recording, or other electronic or mechanical methods, without the prior written permission of the publisher, except in quotations for critical review by reviewers, as permitted by copyright law.

Limits of Liability/Disclaimer of Warranty

The author and publisher of this book and any accompanying material have used their best efforts in preparing this program. The author and publisher make no representation or warranties with respect to the accuracy, applicability, fitness, or completeness of the contents of this program. They disclaim any warranties (expressed or implied), merchantability, or fitness for any particular purpose. The author and publisher shall in no event be held liable for any loss or other damages, including but not limited to special, incidental, consequential, or other damage. As always, the advice of a competent legal, tax, accounting, or other professional should be sought. The author and publisher do not warrant the performance, effectiveness, or applicability of any sites listed in this book. All links are for information purposes only and not warranted for content, accuracy, or any other implied or explicit purpose.

For permissions, contact (Ericka Kelly at https://erickakellyenterprises.com) or the publisher, "Attention: Permissions Desk," at the address below.

MARILEE Publishing
PO Box 238, Altadena, CA 91003-0238
www.marileepublishing.com

ISBN-13 978-1-953569-88-2 (Paperback)
Library of Congress Control Number 2024934975

ISBN-13 978-1-953569-91-2 (Hardback)
Library of Congress Control Number 2024934975

Ordering Information: To place an order, contact (Ericka Kelly, 562-896-9484 / Ericka.kelly@ekenterprises.co) or contact publisher at the address above.

This book does not make any warranties about the completeness, reliability, and accuracy of its information, which reflects the author's present observations and personal or professional experiences over time. Any action taken based on the information found in this book is strictly based on the reader's own beliefs. The publisher and author(s) regret any potential or unintentional harm resulting from this book.

Dedication

To all the military members who gave me the opportunity to have an amazing 32-year career in the Air Force! I want to express my deepest gratitude.

This book is dedicated to my military mentors because, without your belief in me, I would not have achieved all that I have. When you took a chance on me—a shy immigrant young woman with an accent—it changed my life. Your willingness to embrace different perspectives and take risks has shaped my career in ways I never could have imagined.

Your unwavering support and belief in me were the cornerstone of my success. I am forever grateful for the opportunities you have given me, and this dedication is just a small token of my appreciation for everything you have done for me. Your faith in me has profoundly impacted my life.

I am deeply thankful for your wisdom, selflessness, and commitment to showing me a path when I was blind. Each one of you has truly made a difference in my life, and I am honored to dedicate this book to you. Your courage and dedication inspire me every day, and I hope that through this book, I can share the lessons and experiences I learned in the transition process from military to civilian life.

Again, I dedicate this book to you, my mentors, as a reminder of the impact you have had on my life. Your belief in me gave me the strength and determination to succeed, and I want to express my heartfelt appreciation for everything all of you have done. This dedication is a small token of my gratitude for the opportunities you have given me, and I hope this book can honor your investment in me.

Thank you for everything.

Ericka

Acknowledgments

As I sit down after writing this book, I am overwhelmed with gratitude for the countless individuals who have supported and mentored me along the way. To my mentors, coaches, and friends, I extend my heartfelt thanks for recognizing my potential and providing me with opportunities to grow and succeed.

Your guidance, wisdom, and belief in me have pushed me to new heights and encouraged me to challenge myself as an Airman, executive coach, and entrepreneur. The knowledge and skills you've shared with me have shaped my journey and greatly influenced my writing, for which I will forever be grateful.

I also want to express my deep appreciation to the many individuals who have significantly impacted my writing career. From friends who offered invaluable feedback to editors who generously shared their expertise, your contributions have been invaluable. Your honesty, encouragement, and willingness to offer insights have made me a better writer and have contributed to shaping this book into what it is today.

To all the readers who have supported me since my first book was published in 2021, I am truly thankful for your trust, encouragement, and enthusiasm. Your kind words and support for my work have ignited my passion and reaffirmed my purpose. This book is as much yours as it is mine, and I am honored to have you accompany me on this incredible journey.

And to all those who have granted me opportunities, large or small, your impact on my life and career cannot be overstated. This book is about transition and a testament to the power of how support, growth, and kindness from others are necessary for our life journey.

With profound gratitude, I thank all the remarkable individuals who have played a significant role in my growth and success. Your unwavering support and countless opportunities have paved my path, and for that, I am deeply grateful.

Ericka

Table of Contents

Foreword

I tell you from experience! The transition from military to civilian life is a journey filled with unique challenges and opportunities. It requires courage, adaptability, and resilience to navigate this profound shift in identity and purpose. As we stand on the cusp of a new chapter, embracing the changes ahead with open hearts and open minds is essential.

"Take Charge of Your Military Transition" serves as a five-step guide for those who have served and are now embarking on the complex journey of reintegration into civilian society. It offers insights, wisdom, and practical advice on resetting our minds and communication styles and navigating new environments.

Ericka has recognized the unmet needs of our military veterans as they transition out of service and have come together to provide a source of hope, healing, and empowerment. Her experiences, knowledge, and expertise are invaluable resources for anyone seeking to successfully transition from the structured world of the military to the diverse and dynamic landscape of civilian life.

Through the pages of this book, you will find guidance on overcoming programmed beliefs, pivoting by learning a "new language," and embracing emotional and psychological aspects of transition. The personal stories shared within these chapters offer inspiration and affirmation, reminding us that we are not alone in our struggles and triumphs.

As you read this book, please remember to approach your transition with patience, empathy, and a willingness to learn. I sincerely hope this book will help bridge the gap between military and civilian experiences, fostering a deeper understanding and appreciation for your sacrifices and contributions. With deep gratitude for your service and profound respect for your journey, I applaud Ericka for offering this invaluable resource to all of us.

May this book serve as a source of empowerment, compassion, and strength as you embark on this important transition.

Lefford Fate,
Ret Command Chief USAF
Support Services Director, City of Sumter, NC

Introduction

How to Successfully Take Charge
of Your Military Transition

As someone who has been through the challenging process of transitioning from military service to civilian life, I understand the struggle all too well. After 32 years in the Air Force, I found myself at a crossroads, faced with the decision to retire or take on another overseas assignment. As a single mom, I knew it was time to go home and prioritize my family.

The transition process took a backseat as I continued to fulfill my duties until the very last day. I left the Air Force feeling unprepared and unsure of what the future held for me. I missed valuable time with my family and lost thousands of dollars by not taking leave before retirement. My medical records were in disarray, and I did not have the time or energy to address the situation.

However, instead of dwelling on what I had lost, I decided to move forward. I networked tirelessly, determined to start my own coaching business. I took the necessary steps to register my business and began to market coaching, speaking, and training classes. I even wrote a book on military leadership with my friend, retired Air Force Command Chief Lefford Fate titled "P.E.R.S.I.S.T.E.N.C.E.: Lessons from Chiefs Empowering You to Move from Success to Significance as a Transformational Leader."

It was a turning point in my life. I shifted my perspective from feeling sorry for myself to being grateful for the new chapter in my life. I reconnected with my sons and friends, and as I gained credibility as a coach in the corporate world, I found fulfillment in serving transitioning military members to find their path to a new career or start their own business.

I learned a valuable lesson during this time. I had not taken my military transition seriously, and it had cost me dearly. Once I let go of my "hurt feelings," I began to thrive and live out my purpose. I consciously prioritized my personal growth and development to become a better person and a more effective coach. I knew that to help others truly, I needed to be intentional and disciplined in my actions. This meant dedicating daily time to learning, honing my skills, and challenging myself to push beyond my comfort zone.

Today, my sons and I have a solid and beautiful relationship, and I continue to positively impact the lives of transitioning military members. Through this commitment, I have seen my coaching business thrive and found a deeper sense of fulfillment, as people are my life purpose.

The lesson I want to share with all transitioning military members is this: take your transition seriously. It may seem daunting, but with the right mindset and determination, you can turn this change into a new and fulfilling chapter in your life. I encourage you to be proud of what you have accomplished as much as your new journey. By committing to personal growth and development, you can unlock your full potential and make a meaningful impact in your life and those around you. It's not always easy, and there will be challenges along the way, but the rewards of intentional and disciplined growth far outweigh the temporary discomfort.

Overall, Here Are Some General Reasons Why Prepping For Your Transition Is Important:

Emotional

Transitioning from military to civilian life is a significant life change that can come with challenges. Acknowledging and accepting this reality as you prepare to shift is important. This transition period can be an emotional rollercoaster as you leave behind the structure, camaraderie, and routine of military life. It's okay to feel uncertain or apprehensive about what lies ahead.

Options

Reflecting on your skills, strengths, and values can help you identify the type of civilian career that aligns with your personal and professional goals. Researching and exploring civilian career options is essential to finding the right fit for you. Consider the skills and experiences you gained in the military and how they can translate to the civilian workforce. But you can also start from scratch and do something completely new.

Assistance

Attending career fairs, workshops, and networking events can help you connect with civilian employers and learn about different industries and job opportunities. Utilize career counseling services, resume writing workshops, and job search assistance to help you prepare for the civilian job market. These resources can provide valuable guidance and support as you navigate this new chapter of your life.

Mentorship

Depending on your career goals, further education or training may be necessary to enhance your skills and qualifications for the civilian workforce. Consider seeking mentorship and advice from other veterans who have successfully transitioned to civilian careers. Their insights and experiences can be invaluable as you navigate the sometimes-overwhelming transition process.

Programs

Creating a strategic plan for your job search is crucial. Set realistic goals, develop a targeted resume, and prepare for job interviews. Take advantage of transition assistance programs and benefits available to military members separating from service. The VA's Transition Assistance Program (TAP) is designed to help you successfully transition to civilian life by providing resources and support.

This Book Is a Bit Different:

This book will guide you through strategies for cultivating a growth mindset, embracing change, and setting achievable goals for your post-military life. Will delve into learning clear and concise communication styles, especially yours. It will also provide practical strategies for expanding your professional network and cultivating meaningful connections with mentors and coaches. You'll explore techniques for self-discovery, assess your values and passions, and align your career path with your purpose. This book is the ultimate guide to a seamless transition to civilian life.

Maintaining a positive and proactive mindset is key throughout the transition process. Stay open to new opportunities and embrace the potential for growth and success in the civilian workforce. Remember that feeling uncertain sometimes is okay, but staying positive and proactive can help you move forward confidently. You have unique skills and experiences gained in the military, which can be a powerful asset as you embark on your civilian career.

I trust your ability to transform and embrace the process of becoming the best version of yourself. You have the potential to soar and thrive in this new chapter, and I look forward to seeing the incredible impact you will make in the world.

Chapter One

Big Picture Overview

Crush Your Military Transition
Master Your Mindset, Communication, and Commitment

Success and finding your purpose begin with a positive mindset, effective communication, and unwavering commitment. When you embrace these pillars, nothing can stand in your way. These factors are the secret sauce to achieving your goals and living your best life.

Mindset
Read this wisdom from the legendary Zig Ziglar: "Your attitude, not your aptitude, will determine your altitude." It's all about that mindset, baby! Your mindset sets the stage for everything – your success, your relationships, your happiness. So, nurture a positive mindset and watch the magic happen.

Communication
Remember this quote from George Bernard Shaw: "The single biggest problem in communication is the illusion that it has taken place." Isn't that the truth! Effective communication is more than just talking; it's about connecting and understanding

each other. Whether you're chatting with a colleague, a client, or a friend, make sure your message is heard loud and clear.

Commitment

Here's a gem from Vince Lombardi: "The quality of a person's life is in direct proportion to their commitment to excellence, regardless of their chosen field of endeavor." Commitment means sticking with your goals and dreams, even when going is tough. It's about showing up, putting in the work, and never giving up.

Stay committed to your future and goals!

Transitioning from the military to a civilian career or business ownership can be incredibly daunting. Many veterans struggle (I did) to find their place in the civilian world or start and run a business successfully. This process requires a significant shift in mindset, communication style, commitment, strength, and intentional actions. That's why I have put together this step-by-step guide, "Take Charge of Your Military Transition," to help you successfully navigate this pivotal period in your life.

The first main step is to analyze your mindset based on experts' teachings in self-growth and development. Your mindset is crucial in determining your success in your transition. You can make positive changes by reflecting on your current mindset and identifying any negative beliefs or limitations. These teachings from non-military mentors will give you valuable insight into key principles you can apply to your career transition. Creating a vision or goal and staying focused will help reinforce a positive mindset. Seeking mentorship or coaching will also be invaluable in helping you shift your mindset in the right direction.

Understanding your communication style, as outlined in the second main step, is also crucial for a successful transition. Taking a personality and behavior (DISC) assessment, studying the results, and practicing effective communication techniques based on your style will enable you to effectively engage and connect with others in the civilian world or business environment.

Finding commitment and strength to prepare for a new environment, the third main step is setting specific, measurable goals for your transition, seeking support, and developing a detailed plan. This commitment will lay the foundation for your success.

Intentionality is a key theme in the fourth main step. Studying the concept of the terror barrier and the stickman concept will help you navigate the potential challenges or fears you may face. Developing strategies for overcoming the terror barrier, such as visualization or positive self-talk, and seeking accountability partners or mentors will help you step out of your comfort zone and build confidence.

Finding your new life purpose, the fifth main step is a big part of achieving success in your career transition. By reflecting on your talents, passions, and values and setting specific career or business goals aligned with your new purpose, you can pursue education or training opportunities and network to find opportunities that align with your new purpose.

The process outlined in this guide is critical for every reader who wants to successfully transition from the military to a civilian career or own a business. It provides a roadmap to help you shift your mindset, communication style, and commitment and overcome the challenges you will face. By following these steps, you can effectively navigate your transition and find success and fulfillment in your new career or business. So, let's dive in and take charge of your military transition!

Chapter Two

Mind Into Matter

STEP ONE
Harness Your Mindset for a Successful Transition

This Chapter Will Help You:

- Identify and address negative beliefs or limitations holding you back from a successful military transition.

- Learn key principles for success from self-growth and development teachings, such as positive mindset, goal setting, and the power of persistence.

- Create a specific goal and draft a plan for your transition, taking small steps toward your desired outcome.

- Embrace change and approach new opportunities positively, allowing for personal growth and new possibilities.

- Connect with experienced professionals for mentorship or coaching to help shift negative paradigms and provide guidance and support during your transition.

"Whether you think you can or you think you can't,
you're right."
~ Henry Ford

The Harsh Reality of a Neglected Transition:

As a Major Command Chief and the Senior Advisor to the Chief of the Air Force Reserve Component, I was used to a fast-paced work environment. The year was 2017, and retirement was looming just around the corner, only two years away. Mentors had been warning me about preparing for this transition, advising me to take care of myself and start planning for my life post-military, but I was so caught up in my duties that I didn't pay much attention to their words of wisdom.

It wasn't unusual for me to work 14 to 16-hour days, with a day off every two months. I was committed to fulfilling my military responsibilities, and the thought of slowing down or taking a break seemed almost impossible. The mission always came first. In hindsight, I realize now that I should have listened to those mentors and balanced my work with personal care, but I felt I had no choice at the time.

As my retirement date grew closer, I began to feel the pressure mounting. I realized that I had run out of time to adequately take care of everything I needed to do to ensure a smooth transition for myself and my family. It felt as though the military system had left me behind, and I was suddenly faced with the daunting reality that I was not mentally ready to make this significant transition from the military.

The months leading up to my retirement were filled with anxiety and uncertainty. I should have prepared for life after the Air Force and was scrambling to catch up. It took about a year to fix my retirement pay, and instead of being frustrated about the money I had lost during the delay, I chose to reframe my perspective. I told myself it was a gift from me to the Air Force, a final display of dedication and sacrifice.

After this challenging period, I slowly began to find a sense of closure and peace. I refused to dwell on the missed opportunities to prepare for my retirement correctly and instead focused on embracing the future. I realized that I needed to let go of the military mindset that had consumed me for so long. It was time to move on and create a new path for myself.

During this reflection, I decided to start my own coaching business. I wanted to use my experiences and expertise to help others navigate the complex transition from military to civilian life. I also learned the importance of asking for help. I reached out to experts and friends for support, and in doing so, I discovered a newfound sense of community and guidance.

Through this tumultuous journey, I learned a valuable lesson: it is okay to take care of yourself. The military machine keeps going, and as dedicated as we are to our duties, we are all replaceable. I now understand the significance of balance and self-care and strive to impart this wisdom to others facing similar challenges.

As I look back on this transformative period, I am grateful for the lessons learned and the growth that has come from it. My retirement from the Air Force was not just an end but a beginning. It marked the start of a new chapter filled with opportunities, personal growth, and the chance to make a positive impact differently.

Let's Get Started With The Tools You Will Need:

- **Cup of coffee** — Helps create a relaxed and reflective environment for self-analysis and reflection on the current mindset.

- **Quiet place to reflect** — Provides a space for uninterrupted self-reflection and analysis of negative beliefs or limitations.

- **Self-Growth and Development books** — One being "Think and Grow Rich." This offers valuable self-growth and development teachings that can be applied to the career transition, such as a positive mindset, goal setting, and persistence.

- **Journal to practice positive mindset reinforcement** — Allows for reflection on progress and reaffirming goals and aspirations, helping reinforce a positive mindset.

- **Resources for finding mentorship or coaching** — Offers opportunities to connect with experienced professionals who can provide guidance and support in shifting negative paradigms and embracing change.

- **Pen and paper to brainstorm** — Facilitates drafting a plan and outlining the steps needed to achieve specific transition goals.

This chapter is about your mindset and how it can play a critical role in your successful transition out of the military. By the end of this chapter, you will better understand how to analyze your current mindset and apply key self-growth and development teachings to your career transition.

Personal Note: It's absolutely critical for us to understand the significance of transitioning from the 1% that serves in the military to the 99% of the civilian population. It is our responsibility and obligation! When we leave the military, we must shift our mindset, assimilate, and adapt to civilian life, where we will be part of the majority. This transition is no easy feat, but it's necessary for us to integrate into society and thrive outside of the military. We must learn new skills, find employment, and connect with our communities. Understanding the sheer difference in population and mentality is crucial to this process. We must recognize that we are no longer a small percentage of the population and must learn to navigate life among the 99%. By acknowledging this shift and embracing the change, we can successfully transition to civilian life and thrive in our new environment.

Action One: Reflect On Your Current Mindset And Identify Any Negative Beliefs Or Limitations

If you feel negative as you navigate the transition out of the military. It's time to shake off those self-limiting beliefs and take control of your thoughts. Find a quiet spot, grab a cup of coffee, and take a moment to reflect on your current mindset. Are there doubts or fears holding you back? It's important to identify these negative thoughts so you can start working on overcoming them. Settle in and take a deep breath as you challenge those negative beliefs.

Remind yourself of your strengths and past successes, and seek support from friends, family, and fellow veterans who have been through similar transitions. Utilize resources like mentorship programs and counseling services to help process your feelings. It's okay to feel uneasy about the unknown but remember; it's also an opportunity for growth and new experiences. Stay positive, stay focused, and keep moving forward on your journey.

From experience, I'm telling you, it's time to kick those negative thoughts to the curb and step into your power! Find that quiet corner, grab your favorite brew, and tackle those doubts head-on. Take a deep breath and remember all the times you've conquered challenges. You've got the skills, the determination, and the support to thrive in this new chapter of your life.

Don't be afraid to lean on your friends, family, and fellow veterans are there for you every step of the way. Seek mentorship programs, counseling services, and resources to help you process your thoughts and emotions. It's okay to feel uncertain about what lies ahead, but trust that you have what it takes to make it through.

The unknown is scary but also an opportunity for growth and new adventures. Stay positive, stay focused, and keep pressing forward on your journey. You've overcome challenges before, and you'll crush this one too. Keep that coffee in hand and that determination in your heart.

Action Two: Study Self-Growth and Development Teachings To Identify Key Principles To Apply To Your Career Transition

Countless incredible experts have dedicated their lives to studying self-growth and development. One of them is Napoleon Hill. In his book "Think and Grow Rich," Hill shares 13 key principles for success that can be applied to any aspect of life, including your career transition. Look for teachings on positive mindset, goal setting, and the power of persistence. These principles can help you shift your mindset in a positive direction.

Another great resource for this is Brene Brown's book "Daring Greatly: How the Courage to Be Vulnerable Transforms the Way We Live, Love, Parent, and Lead." This has resonated with audiences worldwide. Her insights on embracing vulnerability and building resilience can be incredibly empowering, especially during transition and growth. Brown's insights are eye-opening and can truly reshape how you approach challenges and transitions.

Then there's Simon Sinek, best known for his leadership and motivational speaking. He's about finding your "why" and understanding its impact on personal and professional development. Sinek's teachings can be a game-changer, helping individuals tap into their sense of purpose and drive as they navigate career transitions.

And, of course, we can't forget about Tony Robbins and his book "Awaken the Giant Within: How to Take Immediate Control of Your Mental, Emotional, Physical and Financial Destiny." This guy is a powerhouse in the self-growth and development realm. With his infectious energy and practical strategies for personal and professional success, Robbins has inspired millions to push past limitations and create extraordinary lives. His work is a goldmine for anyone looking to make significant leaps in their career and personal growth journey.

Your military transition is more than a checklist or "just" finding a job. These books might not be what you expected; however, if you do not conquer your thoughts and mindset, your transition will meet tons of resistance and obstacles. Each recommended book helped me move forward; each is a treasure trove of wisdom and inspiration, perfect for anyone seeking to elevate their mindset and navigate career transitions confidently and purposefully.

Action Three: Create A Goal, Draft A Plan, and Act Toward Your Desired Outcome

Once you've absorbed all the wisdom from self-growth and development teachings, it's crucial to translate that knowledge into action. The first step is to set a specific goal for yourself during this transition period. Whether it's a career change, a new hobby, or a personal milestone, having a clear objective in mind will give you a sense of purpose and direction. Next, take the time to draft a detailed plan outlining the specific steps you need to take to achieve this goal. Break it down into smaller, manageable tasks, and assign deadlines to each of them. By acting, no matter how small, you'll begin to build momentum and confidence, paving the way for further progress.

It's normal to feel apprehensive when embarking on a new journey, but remember that every journey begins with a single step. Keep in mind that even the most minor actions can make a big difference. The key is to keep moving forward, no matter how slow or steady the progress may be. You'll be closer to achieving your goal with each step you take. Feel free to seek support from friends, family, or a mentor who can help keep you accountable and motivated. As you see the fruits of your labor and the positive changes in your life, you'll gain the confidence to continue striving for personal growth and development.

Action Four: Be Open To Change And Assimilate To New Environments With A Positive And Healthy Mindset

Transitioning from the military into a new career or entrepreneurship can be exciting and daunting. It's important to approach this change with a positive mindset. Embrace the new opportunities that come your way and be open to learning and growing in the process. Remember that change can lead to exciting new possibilities and that your military experience has equipped you with valuable skills that can be applied in various settings.

As you embark on this new journey, it's important to keep an open mind and be willing to adapt to change. Embracing the transition will help you in your new career or business venture and allow you to continue to thrive and succeed. Remember that you are not starting from scratch - you have a wealth of experience and knowledge that can be applied to your new endeavor. With the right mindset and determination, you can make this transition successful.

An example and recommendation is Earl Nightingale's "The Strangest Secret." It's all about unlocking the key to success and happiness, and it's delivered in a thought-provoking and inspiring manner.

In "The Strangest Secret," Nightingale speaks about how we become what we think about. He emphasizes the power of our thoughts and how they shape our reality. The message is simple yet profound: our inner beliefs and thoughts create the life we lead. It's a call to take control of our minds and focus on positive, empowering thoughts to manifest our greatest desires.

The core idea of the message is to encourage us to shift our mindset and thoughts towards our goals, dreams, and aspirations. "The Strangest Secret" is a timeless classic that inspires and motivates individuals to reach their highest potential. It's an empowering reminder that our thoughts can shape our destinies, and by harnessing that power, we can achieve extraordinary success and fulfillment in life.

Action Five: Seek Mentorship Or Coaching To Help Shift Negative Paradigms

Finding a mentor or coach can make a difference in shifting your mindset. Having someone who can offer guidance, support, and wisdom from their experiences can be incredibly valuable. Look for opportunities to connect with experienced professionals in your field willing to take you under their wing and help you grow. Whether reaching out through online platforms like LinkedIn or attending professional networking events, be pro-active in seeking out those who can offer valuable mentorship.

In addition to seeking a mentor or coach, consider joining a professional organization related to your field. Many of these organizations offer mentorship programs, networking events, and learning opportunities that can all contribute to your mind-set shift. Surrounding yourself with like-minded individuals passionate about personal and professional growth can help inspire and support you. Don't be afraid to put yourself out there and make connections. You never know how these relationships can positively impact your mindset and career trajectory.

Here Are Examples Of How To Find A Mentor Or Coach:

- Before separating from the military, network with professionals in your desired industry or field.

- Identify potential mentors or coaches who have successfully transitioned from military to civilian life.

- Contact individuals who align with your career goals and seek guidance on the transition process.

- Attend professional networking events or workshops to build connections and seek out potential mentors.

- Seek out mentorship programs specifically designed for military members transitioning to civilian careers.

- Visit my website, https://erickakellyenterprises.com/ and participate in my monthly complimentary leadership military conversations

As you work through these action steps, remember to take the time to journal and practice reinforcing a positive mindset. Use your journal to reflect on your progress and reaffirm your goals and aspirations. By analyzing your current mindset, studying key teachings, setting goals, acting, being open to change, and seeking guidance, you'll be well on your way to harnessing your mindset for a successful transition out of the military. Keep up the great work and remember that a positive mindset can make all the difference!

Chapter Summary:

- How to identify and overcome negative beliefs or self-limitations.

- Read amazing books where you can learn positive mindset, goal setting, and the power of persistence.

- Create specific goals for your career transition.

- Be open to change and assimilate into new environments.

- Take small steps towards your goal and build momentum and confidence.

Chapter FAQs:

How can I identify negative beliefs or limitations in my current mindset?

You can identify negative beliefs or limitations in your current mindset by finding a quiet place to reflect and relax. Think about any doubts or fears about your skills or the transition out of the military. These negative thoughts can hold you back, so it's important to identify them to overcome them.

What fundamental principles for success can I learn from self-growth and development teachings?

One excellent resource for critical principles for success is Napoleon Hill's book "Think and Grow Rich." In this book, you can learn about a positive mindset, goal setting, and the power of persistence. These teachings can help you shift your mindset in a positive direction and apply them to your career transition.

How can I create a specific goal for my career transition?

You can create a specific goal for your career transition by identifying what you want to achieve. For example, if you want to start a new business, your goal might be to launch the business within six months. Having a clear and specific goal will give you direction and purpose as you work through your transition.

What are some ways to be open to change and assimilate into new environments with a positive mindset?

To be open to change and assimilate into new environments with a positive mindset, embracing new opportunities, and being open to learning and growing is important. You can approach change with curiosity and excitement, knowing it can lead to new possibilities and experiences.

How can I find a mentor or coach to help shift negative paradigms in my mindset?

If not me, you can find a mentor or coach by networking with experienced professionals in your field. Online platforms like LinkedIn or professional networking events can be great places to connect with potential mentors. Contact individuals whose experiences and knowledge align with your career goals and seek their guidance and support.

How can I reinforce a positive mindset through journaling?

You can reinforce a positive mindset through journaling by regularly reflecting on your progress and reaffirming your goals and aspirations. Use your journal to track your journey, celebrate your wins, and remind yourself of the positive changes you are making in your mindset and career transition.

What are the benefits of seeking mentorship or coaching in shifting negative paradigms?

Seeking out mentorship or coaching can provide guidance, support, and strategies for shifting negative paradigms in your mindset. A mentor or coach can offer valuable insights, share their experiences, and help you navigate the challenges of transitioning out of the military, ultimately helping you develop a more positive and growth-oriented mindset.

What are some common negative beliefs that veterans may hold about their skills or the transition out of the military?

Veterans may hold opposing beliefs, such as feeling unqualified for civilian jobs, doubting their transferable skills, or fearing the unknown aspects of life outside the military. Overcoming these negative beliefs is crucial for a successful transition, and seeking out mentorship or coaching can be valuable in addressing and overcoming these beliefs.

How can a positive mindset and goal setting help in a successful transition out of the military?

A positive mindset and goal setting are essential in a successful transition out of the military because they provide clarity, purpose, and motivation. By shifting your mindset and setting specific goals, you can stay focused, overcome challenges, and achieve meaningful results in your career transition.

How can taking small steps toward my goal help build momentum and confidence in my career transition?

Taking small steps towards your goal can build momentum and confidence by showing progress and providing a sense of accomplishment. Each small step reinforces the belief that your goals are attainable, boosting your confidence and motivation to progress toward a successful transition.

Chapter Three

Communication Mastery

STEP TWO
Leverage Your Personality and Behavior (DISC) Profile for Success

This Chapter Will Help You:

- Identify your communication style through the DISC assessment.

- Understand your strengths and weaknesses in communication.

- Practice effective communication techniques based on your style.

- Seek out opportunities to practice and refine your communication skills.

- Use your newfound understanding to engage and connect with others effectively.

"In a very real sense we have two minds,
one that thinks and one that feels."
~ Daniel Goleman

The Shocking Moment That Made Me Rethink Everything:

I remember that day like it was yesterday, when my son, Sean, at only eight years old, called me out on my behavior. It was a typical day; I was on the phone with one of my subordinates in the Air Force when Sean came home from school and tried to show me something. I was so deep into my conversation that I ignored him, giving him looks I now regret. I was so focused on being professional and in control that I didn't even notice the hurt in his eyes.

After the call, I lashed out at Sean, scolding him for interrupting me. I lectured him about breaking the rules and being impolite. I didn't stop to think about the impact of my words on him. He stood there, trembling with tears in his eyes, and asked me the question that would change everything: "When are you going to like me as much as you like your military friends?"

Those words hit me like a punch in the gut. I had been so focused on being the best Airman I could be that I had neglected my children. At that moment, I realized the toll my dedication to the military had taken on my family.

I held Sean in silence, overwhelmed with the realization that I needed to change. I needed to learn to connect better with my children and treat them with the same love and respect I showed to my comrades in the Air Force. It wasn't just about being a better Airman; it was about being a better person and parent.

From that day on, I made a conscious effort to prioritize my sons and to be present in their lives. I learned to balance my dedication to my career with my responsibilities as a mother. I made time for family dinners, weekend outings, and heartfelt conversations with my boys. I learned to listen to their stories, support their dreams, and be there for them in a way I had never been before.

As I transitioned from the military into a new career, I carried lessons from that transformative moment with my son. I approached my new path with a newfound sense of balance, understanding that success in my career should never come at the cost of my relationships with those I love. I became a mentor to transitioning military members, using my own experiences to guide and support them as they navigated the challenges of leaving the service.

In the end, Sean's brave question sparked a profound change within me. I had grown into a stronger, more compassionate person, capable of connecting with others on a deeper level. I had become the kind of leader who valued professional success and the richness of human relationships and connections. As I embarked on this new chapter in my life, I did so with confidence, knowing that I was not just a better Airman but a better person because of the love and lessons of my son.

For This Chapter, I Invite You To:

- Get a DISC assessment done. This chapter is critical because it helps you understand your communication style and how to connect with others effectively.

- Spend time learning about your emotional intelligence, as it is essential in this chapter because it provides valuable knowledge on navigating and understanding emotions in oneself and others, which is crucial for effective communication.

- Please continue to read self-development books. Journaling is important in this chapter because it helps track progress and further develop communication skills.

- Watching your positive mindset is critical in this chapter because it helps maintain a confident and optimistic attitude when practicing and refining communication skills.

- Finding mentors or an accredited coach is essential in this chapter because they can provide guidance and support as the transitioning military members develop their communication skills and confidently transition into a new career or entrepreneurship.

Communication is a critical skill that can help you succeed in your transition from the military into a new career or entrepreneurship. This chapter will focus on understanding and leveraging your communication style to connect with others effectively.

As a DISC trainer and consultant, I can tell you a DISC assessment is the ultimate way to decode human behavior; it is a powerhouse. Here is a quick overview:

D — Dominance. These are your go-getters, the take-charge types who set their eyes on the prize and won't stop until they get it. They're the alpha dogs, always looking to conquer new challenges. You want to communicate with them straight to the point, no fluffy stuff; get to the point and watch them jump to action.

I — Influence. These are individuals who thrive on connections and relationships. The social butterflies with charisma for days can charm the pants off anyone. When you're chatting with I's, make it engaging, add some pizzazz, and get chatty. They feed off enthusiasm and love to be the life of the party.

S — Steadiness. These are your dependable, supportive peeps who value harmony and teamwork. They're the glue that holds everything together. When communicating with Ss, it's all about empathy, support, and building trust. Keep it calm, avoid big changes, and hold their hand through the process.

Last but not least, **C — Conscientiousness.** They represent the logical, detail-oriented folks. Precision is their middle name! They're all about accuracy, data, and getting it right. When talking to the C's, prepare to be thorough and factual. Show them the evidence, cross your T's, dot your I's, and you'll earn their respect.

Understanding your DISC profile is like having a secret code to unlock the mysteries of human interaction! Knowing someone's DISC style allows you to tailor your communication to speak their language, build rapport, and avoid misunderstandings.

Personal Note: As a single mom with two teenage sons, it seemed like there was constant tension and miscommunication in our household. I didn't understand why my strict, detail-oriented nature clashed with their dominant and supportive personalities. It wasn't until I became a DISC consultant and we took the assessments that everything began to change. I discovered that I was a high C style, which explained why I was so focused on rules and order. On the other hand, my sons were a combination of D and S, which meant they were driven and helpful. understanding these differences and learning how to communicate effectively with each other completely saved our relationship. Instead of constant arguments, we now have respectful and productive conversations, and we've learned to appreciate each other's unique strengths. Learning about our DISC styles improved our relationships and helped me become a better mom and leader for my sons. I now understand why my need for structure and organization can be overwhelming for my sons, and I've learned to balance that with their need for independence and freedom.

Action One: Take A DISC Assessment To Determine Your Communication Style

When you take a personality and behavior assessment like DISC, you're not just learning about how you communicate - you're gaining a deeper understanding of yourself. By recognizing your dominant communication style, you can begin to embrace your strengths and work on any areas that may need improvement. It's a foundational step to becoming a more effective and confident communicator.

The results of a DISC assessment can also shed light on how you prefer to receive communication from others. This self-awareness is essential for building strong and successful personal and professional relationships. It can help you adapt your communication style to better connect with others and ensure your message is received as intended.

Utilizing the insights from a DISC assessment can also help you become more empathetic and considerate in your interactions. By understanding that not everyone communicates the same way you do, you can tailor your approach to meet the needs of others. This leads to more productive and harmonious relationships and greater understanding and respect within your personal and professional circles.

Taking a DISC assessment is a proactive and empowering step towards honing your communication skills. It provides a solid foundation for self-improvement and growth, enhancing communication, understanding, and relating to others. With this newfound knowledge, you can approach your interactions confidently and clearly, creating positive and meaningful connections with those you encounter.

Action Two: Study The Results And Identify Your Strengths And Weaknesses In Communication

After completing the assessment, diving into the results and recognizing your communication strengths and weaknesses is crucial. Imagine if you lean towards a dominant communication style – you might thrive in leadership roles but encounter difficulties with empathetic communication. Pinpointing these areas of proficiency and those needing improvement will guide you toward targeted growth.

When reviewing your results, note recurring patterns in your communication style. Reflect on whether certain people or situations present communication challenges for you. Are there consistent hurdles in effectively conveying your thoughts and ideas? These insights will allow you to identify areas to refine and develop your communication skills to elevate your interactions.

It's also important to recognize and celebrate your strengths in communication. You could listen actively and demonstrate empathy or excel at presenting your ideas clearly and persuasively. You can leverage these strengths to enhance your communication effectiveness by acknowledging and appreciating them.

Once you clearly understand your strengths and weaknesses, plan to improve your communication skills and how you connect with others. This might involve seeking out training or resources to help you develop in areas where you could be more confident, such as practicing active listening or learning to assert your ideas more assertively. Additionally, consider finding a mentor or coach who can provide guidance and support as you work to enhance your communication abilities.

Understanding your strengths, fears, and challenges in communication is an important step in personal and professional growth. By studying your results and developing an improvement plan, you can become a more effective and confident communicator, leading to stronger relationships and tremendous success in all areas of your life.

Action Three: Practice Effective Communication Techniques Based On Your Primary Style

When communicating as a friendly and inspirational writer, your interactions must be warm and approachable. This means using positive and encouraging language and focusing on connecting with your audience. You can practice this by using storytelling and personal anecdotes to inspire and engage your readers while providing precise and actionable advice.

In addition to being friendly and inspirational, it's important to be instructional in your communication style. This means providing clear and practical guidance that your audience can follow. To practice this, focus on breaking down complex ideas into simple, easy-to-understand steps. Use examples and visuals to illustrate your points and make it easier for your audience to absorb and apply the information you share.

Another effective communication technique for a friendly and inspirational style is empathy and understanding. This means acknowledging the emotions and concerns of your audience and offering support and reassurance. Practice active listening and responding with empathy, showing that you genuinely understand and care about the needs and challenges of your readers.

An essential practice for effective communication is to be genuine and authentic in your communication. People are drawn to those who are transparent and honest, so it's important to be yourself and let your personality shine through. Practice this by connecting from the heart and sharing your experiences and lessons learned. By being genuine, you can build trust and credibility with people and create a deeper connection that will resonate with them.

Action Four: Seek Out Opportunities To Refine Your Communication Skills

Seeking opportunities to practice and refine your communication skills is critical for personal and professional growth. Whether engaging in thoughtful conversations with colleagues, participating in public speaking events, or even just chatting with friends, every interaction provides a chance to hone your communication abilities. By actively seeking out these opportunities, you demonstrate an eagerness to improve and a willingness to step out of your comfort zone.

Effective communication is not just about what you say but also about how you convey your message. Paying attention to your body language, tone of voice, and writing style is important. Practicing in different settings allows you to experiment with various communication styles and techniques, ultimately helping you to become a more versatile and impactful connector.

Networking events are an excellent place to test and refine your communication skills. Whether participating in a casual conversation or presenting your ideas to a group, networking events provide a valuable opportunity to practice speaking and listening effectively. Additionally, team meetings in the workplace can allow you to collaborate with colleagues and exercise your communication skills in a professional setting.

Even casual conversations with family and friends can serve as an excellent platform for practicing communication. Engaging in meaningful discussions and listening to others can help you become more empathetic and articulate in your communications. Effective communication involves getting your point across and understanding and connecting with others. By seeking opportunities to practice and refine your communication skills, you can become a more confident and influential communicator in all aspects of your life.

Action Five: Use Your Newfound Understanding To Engage And Connect With Others Effectively

When engaging with others, show genuine interest and active listening. Ask open-ended questions to encourage them to share their thoughts and experiences. Demonstrating empathy and understanding can create a deeper connection with the person you are communicating with. This will help you build solid relationships and enable you to effectively collaborate and work with others towards shared goals.

Additionally, be mindful of nonverbal communication. Your body language, facial expressions, and tone of voice can all impact how effectively you connect with others. Practice maintaining open and approachable body language, making eye contact, and using a warm and friendly tone. These subtle cues can convey your sincerity and interest in the conversation, making it easier for others to engage with you.

Furthermore, be adaptable in your communication style. Recognize that different individuals may respond better to certain communication styles or approaches. By being flexible and adjusting your communication style to suit the person you are engaging with better, you can create a more comfortable and productive interaction. Remember, the goal is to create a positive and meaningful connection with others, and adaptability is critical to achieving this.

Always strive to leave a positive impression. Whether through a friendly smile, a genuine compliment, or a thoughtful follow-up after a conversation, aim to build a positive rapport with those you engage with. Showing genuine care and consideration for others will enhance your ability to connect effectively and leave a lasting impact on those you interact with. Using your newfound understanding and skills, you can engage and connect with others more effectively and create meaningful relationships that benefit you personally and professionally.

Recommended Resources:

To help you along the way, consider investing in a personality and behavior assessment to gain further insight into your communication style. Additionally, learning about emotional intelligence can provide valuable knowledge on navigating and understanding emotions in yourself and others. Continue to read self-development books and journal your experiences to track your progress. Lastly, seek mentors or an accredited coach who can guide and support you as you develop your communication skills.

By following these steps and utilizing the recommended resources, you can master communication and confidently transition into a new career or entrepreneurship.

Chapter Summary:

- You can effectively assess your communication style by taking a DISC assessment.

- It is essential to identify your strengths and weaknesses in communication.

- Build rapport, create enthusiasm, and focus on empathy.

- Practice and refine your communication skills.

- Investing in an emotional intelligence assessment.

- Leverage your newfound understanding to engage and connect with others effectively.

- Convey your ideas and goals effectively.

Chapter FAQs:

How can I effectively determine my communication style?

You can effectively assess your communication style by taking a DISC assessment. This assessment will help you identify whether you are Dominant, Influential, Steady, or Conscientious in your communication style.

Why is it important to identify my strengths and weaknesses in communication?

Identifying your strengths and weaknesses in communication is essential because it will help you focus on areas for improvement. For example, if you lean towards being a Dominant communicator, you may excel in leadership roles but need help with empathetic communication.

What are some effective communication techniques based on my style?

Some effective communication techniques based on your style may include building rapport, creating enthusiasm, or focusing on empathy, depending on your communication style. For example, if you are an Influential communicator, building rapport and creating enthusiasm in your interactions may be effective techniques.

How can I practice and refine my communication skills?

You can practice and refine your communication skills by seeking opportunities in different settings, such as networking events, team meetings, or casual conversations. The more you practice, the more confident and skilled you will become in your communication.

How can I leverage my newfound understanding of my communication style to engage and connect with others effectively?

You can leverage your newfound knowledge of your communication style to engage and connect with others effectively by adapting your communication style to communicate with different personalities and build strong relationships. This can be done in job interviews, business meetings, networking events, and other interactions.

What resources are recommended for further insight into my communication style?

Some recommended resources for further insight into your communication style include investing in a personality and behavior assessment, learning about emotional intelligence to navigate and understand emotions in yourself and others, reading self-development books, journaling your experiences to track your progress, and seeking out mentors or an accredited coach for guidance and support.

How can I confidently transition into a new career or entrepreneurship by mastering communication?

You can confidently transition into a new career or entrepreneurship by mastering your communication through understanding your communication style, practicing effective communication techniques, seeking opportunities to practice and refine your skills, and leveraging your newfound understanding to engage and connect with others effectively.

Why is effective communication important in transitioning from the military to a new career or entrepreneurship?

Effective communication is important in transitioning from the military to a new career or entrepreneurship because it is a critical skill that can help you succeed in your new endeavors. It allows you to connect with others, build relationships, and effectively navigate different interactions and settings.

How can I effectively adapt my communication style to connect with different personalities?

You can effectively adjust your communication style to different personalities by leveraging your understanding of emotional intelligence, practicing empathy, and using techniques that align with the other person's communication style. For example, for someone with a Steady communication style, you may focus on being patient, supportive, and collaborative in your interactions.

What are the benefits of mastering my communication skills during transition?

The benefits of mastering your communication skills in the transition process include building strong relationships, creating opportunities for career advancement, effectively conveying your ideas and goals, and ultimately succeeding in your new career or entrepreneurial endeavors. Effective communication can also lead to better teamwork and collaboration.

Chapter Four

Commitment to Success

STEP THREE
Navigate the New Terrain

This Chapter Will Help You:

- To stay motivated and determined, understand the importance of reflecting on your reasons for transitioning from the military into a new career or business ownership.

- Set specific, measurable goals for your transition, allowing you to track your progress and maintain focus on what you want to achieve.

- Seek support from fellow veterans or business owners who have made similar transitions, as they can provide valuable insights, advice, and encouragement.

- Develop a detailed plan for your career or business transition, breaking your goals into actionable steps and creating a timeline for achieving them.

- Commit to taking consistent action steps towards your new environment by staying proactive and persistent in pursuing your goals.

"Success is not just about reaching the destination but navigating the new terrain. It's about embracing challenges, adapting to change, and staying committed to your goals, no matter what obstacles may come your way. Remember, success is not a straight path but a winding road filled with twists and turns. Stay dedicated, stay resilient, and keep moving forward. You've got this!"
~Anonymous

Unbroken Wounds: The Inspiring Journey Of A Retiring Soldier's Transformation

Army Sgt Maj. Jones stood on the threshold of a new chapter in his life, the summer of 2020 marking the end of his 30 years of dedicated service. As he faced the daunting transition into civilian life, he was engulfed in a tangle of emotions, fears, and unresolved issues that had lingered in the shadows for far too long.

Throughout his military career, Jones had poured his heart and soul into his work, leaving his wife and children on the periphery of his existence. His free time was consumed by the camaraderie of his Army friends, who understood and shared the burden of multiple deployments and the heart-wrenching loss of comrades on the battlefield. He had retreated into himself, denying the mounting mental health issues and suffering in stoic silence, unable to confront the demons that plagued his mind and soul.

As the prospect of retirement loomed ever closer, Jones found himself unequipped to make the crucial decisions that would shape his and his family's futures. His relationship with his college-age son had fractured, and the specter of civilian life gripped him with an overwhelming sense of uncertainty. With only a high school diploma and a skill set predominantly rooted in combat, the prospect of finding a viable job seemed bleak and remote, casting a shadow of despair over his hopes for a fulfilling and satisfying future.

But as the relentless tide of change bore down upon him, Jones found a glimmer of hope in the form of a coach who guided him through the turbulent waters of transition. He began to recognize the importance of including his wife in the plans for their new life, seeking to bridge the chasm that had grown between them. Jones enrolled in the same college classes to repair the fractured bond with his son, determined to forge a connection by walking alongside him through this pivotal time in their lives.

With the newfound determination to confront the challenges ahead, Jones defied the shackles of pride and sought the help he desperately needed. He delved into the world of transitional conversations beyond the military sphere. He sought to nurture his mental health, recognizing that the scars of war ran more profound than the physical wounds he had sustained. These pivotal steps laid the foundation for a metamorphosis that began to unfold before his eyes.

Jones found his footing again as the days unfurled into weeks, and the weeks melded into months. He secured a local store manager job, drawing upon his unwavering discipline and leadership skills to carve out a new path for himself. With diligence and patience, he began to nurture the bonds that had frayed in the crucible of his military service, learning to smile, dream, and imagine a future illuminated by hope and positivity.

The tumultuous journey of retirement had forced Jones to confront his demons, but in doing so, he unearthed the strength to mend his marriage and bridge the divide with his son. Through the tumultuous storm of change, he emerged not as a defeated warrior but a man rekindled with purpose and a vision for a brighter tomorrow. He had learned to embrace vulnerability and seek help when needed, no longer burdened by the weight of silence that had once suffocated his spirit.

As the curtain closed on his military career, Jones found himself standing on the precipice of a new beginning, with the love of his family by his side and the promise of an unwritten future awaiting him. With a fiery determination and a tender heart, he stepped into the unknown, ready to embrace the boundless possibilities ahead.

For This Message, You Need To:

- Control your actions throughout the transitional process.

- Keep learning about your mindset.

- Continue to read self-development books.

- Find mentors or an accredited coach.

Personal Note: Transitioning from the structure and certainty of the military to the unpredictability and risk of starting a business was a daunting challenge. I had to be intentional and decisive in my actions, relying on my determination to guide me through the unknown. I knew that I wanted to become an accredited coach and positively impact people's lives, but I had no experience as an entrepreneur. Despite the uncertainty, I knew that I had to take the leap and pursue my passion. As I embarked on my entrepreneurial journey, I quickly realized that it would require immense dedication, time, and financial investment. There were moments of doubt and fear, but I remained steadfast in my purpose, determined to navigate through the obstacles. The process of building my coaching business tested my resilience and forced me to adapt to new challenges. Despite the difficulties, I found immense fulfillment in my progress and the impact I could have on others. Looking back, I am grateful for the opportunity to pursue my passion and build something meaningful from the ground up.

Action One: Staying in Control and Focused Means Being Able to Manage Your Emotions and Reactions

It's vital to maintain a sense of calm and composure, even when faced with difficult or unexpected situations. This might mean taking a step back, assessing the situation before reacting, or simply reminding yourself of your ultimate goals and priorities.

By staying focused on the bigger picture, you'll be better equipped to handle any obstacles that come your way.

Another critical aspect of staying in control and focused is prioritizing your tasks and responsibilities. This might involve creating a schedule or to-do list to help you stay organized and on track. By breaking down your goals into smaller, manageable steps, you can tackle them one at a time, which can help prevent feeling overwhelmed or stressed. It's also important to set boundaries and say no to things that might distract you from your goals to maintain focus and stay on course.

In addition, maintaining a positive attitude is crucial when it comes to staying in control and focused. Believing in yourself and your abilities will make you more likely to stay motivated and on track. It's also important to surround yourself with a supportive and uplifting community, whether friends, family, or mentors, who can help keep you in the right frame of mind. By maintaining a positive attitude, you'll be better equipped to face challenges and stay focused on your goals, even when things get tough.

Staying in control and focus is about being proactive and intentional with your actions. It's about taking ownership of your choices and staying dedicated to your goals, no matter what external factors might come into play. By staying disciplined and committed to your vision, you can confidently and confidently navigate any transitions or changes.

Action Two: Consider The Importance Of Continual Learning And Growth In The Field Of Self-Development

It's easy to become complacent once we feel comfortable in our current roles or positions, but the truth is that there's always room for improvement. Self-help books offer a wealth of knowledge and strategies to help us better navigate life, career, and business endeavors. We can glean valuable insights and perspectives to enhance our personal and professional growth by dedicating time to reading and learning from these resources.

Another critical point to emphasize is the impact of writing style in self-help literature. These books' friendly and relatable tone encourages readers to engage with the material personally. When the writing style is inspirational and instructional, it creates a sense of motivation and empowerment, urging readers to apply the concepts and strategies to their lives. By adopting a friendly and approachable tone, authors can effectively connect with readers and inspire them to take positive action toward self-improvement.

Moreover, the content found in self-help books often serves as a guiding light for individuals navigating new career paths or venturing into entrepreneurship. Whether it's advice on building confidence, networking, time management, or leadership skills, these resources offer practical tools and techniques to help readers succeed in their professional endeavors. By continuously reading up on self-development, individuals can stay informed about the latest trends and best practices in their respective fields while gaining valuable insights into personal growth and well-being.

Self-help books are invaluable resources that provide a wealth of knowledge and guidance for personal and professional development. Individuals must prioritize continual learning and growth by consistently engaging with these motivational and instructional materials. By doing so, they can glean new insights, adopt empowering strategies, and enhance their overall well-being and success in life and work.

Action Three: Don't Go Alone When Facing A Significant Life Transition Or Challenge!

Finding yourself a mentor or coach can be a game-changer. Having someone in your corner who can offer advice, encouragement, and insights can make all the difference in staying on track and achieving your goals. A mentor can provide valuable guidance based on their own experiences and help you navigate the ups and downs of your journey.

It's important to remember that seeking guidance from a mentor doesn't mean you're weak or incapable. It's a sign of strength and self-awareness to recognize when you could benefit from the support of someone who has been where you are and can offer valuable perspective. A mentor can help you see things from a different angle, challenge your limiting beliefs, and push you to stretch beyond your comfort zone.

In addition to providing emotional support and guidance, a mentor can help you develop specific skills and strategies to achieve your goals. Whether in professional development, personal growth, or navigating a new phase of life, a mentor can provide a roadmap for success. They can share their knowledge, resources, and network to help you make informed decisions and avoid common pitfalls.

If you feel overwhelmed or unsure of your next steps, seek a mentor or coach to guide you. Remember, you don't have to go it alone. With the proper support and guidance, you can stay focused and motivated and ultimately crush your goals.

Chapter Summary:

- Reflect on your reasons for transitioning and set specific, measurable goals.

- Seek support from fellow veterans or business owners who have made similar transitions.

- Develop a detailed plan for your career or business transition.

- Commit to taking consistent action steps towards your new environment.

- Use suggested resources and tools to help navigate the transition, stay disciplined, and focused on your goals.

Chapter FAQs:

What specific steps can be taken to find commitment and strength to prepare for a new environment?

Some specific steps to find commitment and strength for a new environment include reflecting on your reasons for making the transition, setting clear and measurable goals, seeking support from fellow veterans or business owners, developing a detailed plan for the transition, and committing to taking consistent action steps towards your goals.

Why is understanding your "why" before navigating a new terrain important?

Understanding your "why" is important because it gives you the motivation and determination to navigate the new terrain ahead. Knowing your reasons for the transition will help keep you focused and driven in your pursuit of success in the new environment.

How can one seek support from fellow veterans or business owners during the transition?

To seek support from fellow veterans or business owners, you can join networking groups or organizations expressly for veterans transitioning into new careers or business ownership. Additionally, you can reach out to individuals who have already made similar transitions for insights, advice, and encouragement.

What suggested resources and tools to consider for navigating the new terrain?

Some suggested resources and tools for navigating the new terrain include self-development books, finding mentors or an accredited coach, attending seminars or workshops, and upgrading your skills. These resources can provide valuable insights, guidance, and support as you navigate the challenges of transitioning.

Why is it crucial to have a detailed career or business transition plan?

A detailed plan is vital because it provides a roadmap and helps keep you on track toward your goals. Breaking down your goals into actionable steps and creating a timeline for achieving them helps ensure that you make progress in your transition.

What actionable steps can be taken toward a new environment?

Some actionable steps towards a new environment may involve networking, attending seminars or workshops, upgrading your skills, and staying disciplined and focused on your goals. These actions can help you progress and prepare for success in the new environment.

How can one stay proactive and persistent in pursuing their goals during the transition?

To stay proactive and persistent in pursuing your goals, you can set regular milestones to track your progress, remain disciplined in your actions, and seek continuous learning opportunities to enhance your skills and knowledge in your new environment.

How do you stay disciplined and focused on your goals during the transitional process?

Some ways to stay disciplined and focused on your goals include creating a daily or weekly schedule, setting reminders for important tasks, and staying committed to your plan even when faced with challenges or setbacks.

Why is it important to keep learning about your mindset and how it can impact your success?

It is essential to keep learning about your mindset because a positive and growth-oriented mindset can significantly impact your success in a new environment. Understanding how your mindset influences your actions and decisions can help you overcome obstacles and maintain motivation.

How can one develop the strength and capability to successfully transition from the military into a new career or business ownership?

One can develop the strength and capability to successfully navigate the transition by staying focused on their goals, seeking support when needed, and being proactive in taking steps toward their new environment. Additionally, staying committed to their plan and continuously learning and improving will contribute to their success in the transition.

Chapter Five

Harness Intentionality

STEP FOUR
A Tool for Pushing Through the Terror Barrier

This Chapter Will Help You:

- Understand the concept of the terror barrier and how it relates to your career transition.

- Identify potential challenges or fears you may face in your transition, such as fear of failure, the unknown, or not being good enough in a new career.

- Develop strategies for overcoming the terror barrier, such as visualization and positive self-talk.

- Practice stepping out of your comfort zone in small ways to build confidence and prepare for bigger leaps.

- Seek accountability partners or mentors who can provide support, guidance, and encouragement as you navigate the terror barrier.

"Make your move before you're ready."
~ *Price Pritchett, PhD*

From Pentagon to Life Purpose: How I Found Fulfillment After Retirement

As I approached my retirement from the Air Force in April 2019, I was filled with a mix of emotions. On the one hand, I was excited to start a new chapter in my life, but on the other hand, I felt a sense of uncertainty and apprehension about what the future held for me. I had dedicated so much of my life to the military, and transitioning into a civilian career or entrepreneurship was thrilling and daunting.

I couldn't shake the feeling that my upcoming civilian job did not give me the same sense of purpose and fulfillment that I had found in the military. My accomplishments and skills in the Air Force were not fully recognized, and I was told to "do my time" rather than being valued for what I could bring.

I also faced the daunting prospect of leaving the Pentagon and returning to Southern California. My civilian job was scheduled to start immediately after my arrival, leaving me with no time to process, reset, or transition. With my mental health issues being treated by military experts, I knew that I faced the challenge of transitioning to the VA system and dealing with the complexities that would come with that.

As I made the cross-country move, I realized I hadn't given myself the time to take care of myself. I had made the mistake of putting my mental health treatment on hold, and as a result, I felt isolated and alone in a new environment. My family was in California, but it felt like the military was leaving me behind, and I struggled to find my place in this new phase of life.

The drive from the East Coast to California allowed me to reflect on my goals and not see my current situation as a reflection of my future. As I began my civilian job, I consciously decided to reset my mindset and not let the negative energy in my new environment bring me down. I was grateful to be there and eager to make a positive impact.

However, one of the most complicated steps I took was to go to the VA and begin my mental and physical health journey. It was a complex and emotional process, but with the support of my family, friends, and other veterans, I began to feel less alone and more hopeful about the future.

I realized that my experiences and struggles had equipped me with the unique ability to empathize and relate to other transitioning military members. I decided to use my lessons to help others in their journey. I began to provide coaching, mentoring, and training to support fellow veterans as they navigated their transitions. I also pursued additional educational certifications that allowed me to realize my true purpose of helping others build their leadership, communication, and influence skills.

Throughout this challenging time, I was fortunate to have a fantastic network of military friends and mentors who lifted me and encouraged me to keep going, even when the road ahead seemed daunting. In the end, I found that by taking the time to prioritize my well-being and seek support from those around me, I emerged from this transitional period with a renewed sense of purpose and deep gratitude for the journey that had brought me to where I was.

Some Recommended Tools You Will Need:

Books or articles on the concept of the terror barrier and how it relates to career transitions are critical to helping the transitioning military members understand the concept of the terror barrier and how it specifically applies to their career transition.

Visualization apps or techniques for overcoming the terror barrier are essential for helping the group push through their fears and uncertainties by visualizing themselves successfully navigating their career transition.

Confidence-building exercises and self-help resources are crucial for helping transition military members step out of their comfort zones and build the confidence needed to overcome the terror barrier.

Online platforms for finding accountability partners or mentors are critical for providing support, guidance, and encouragement as the members navigate the terror barrier and make their career transition.

In this chapter, we will dive into the concept of the terror barrier and how it relates to your career transition. The terror barrier is the feeling of fear and uncertainty that holds us back from making a change or taking a big step. That voice in our head says, "You can't do it," or "What if you fail?"

Personal Note: Facing fear has been a constant in my life from a very young age. As a child, I survived through abuse, hunger, and various diseases in Central America. These experiences taught me the resilience and courage I needed to face the many challenges life threw me. When I came to the US, I faced language and cultural barriers that initially seemed insurmountable. However, I refused to let fear hold me back from experiencing all that life had to offer, and I pushed through those barriers with determination and a strong spirit. In my early adulthood, I found myself in a very violent relationship, and I had to summon every ounce of strength and courage to break free from it. Once in the Air Force, I was confronted with gender and racial biases that made me feel like an outsider in my own workplace. Despite these adversities, I continued to fight through the terror barrier, refusing to be limited by fear. My life's journey has been one of facing fear head-on, and I am proud of the resilience and bravery I have shown in the face of adversity. Through sharing my experiences, I hope to inspire others to confront their own fears and never allow them to hold them back from living their best lives.

The terror barrier is a psychological concept that refers to a mental block or obstacle preventing us from moving forward. It is often associated with fear, self-doubt, and limiting beliefs that hold us back from reaching our full potential. The terror barrier can come from past experiences, negative conditioning, and societal expectations that cause us to feel stuck and unable to progress.

The terror barrier may have been formed in childhood through experiences of failure, rejection, or trauma. Over time, these negative experiences become ingrained in our subconscious and create a sense of fear and resistance towards taking risks or pursuing our goals. As a result, we may feel paralyzed by self-doubt and unable to move past the barriers that prevent us from achieving our dreams.

Overcoming the terror barrier requires challenging and reprogramming our limiting beliefs and fears. It involves stepping out of our comfort zone, taking risks, and confronting the negative thoughts that hold us back. By building resilience, self-confidence, and a positive mindset, we can gradually break through the terror barrier and create new possibilities for personal growth and success.

It's important to remember that the terror barrier is not permanent and can be overcome with determination and a willingness to change our mindset. By accepting that fear and self-doubt are natural parts of the human experience, we can begin to shift our focus toward growth and transformation. With perseverance and the right support, we can break through the terror barrier and pave the way for a more fulfilling and empowered life.

Action One: The Concept Of The Terror Barrier And How It Relates To Your Career Transition

Understanding the terror barrier and how it might appear in your life is the first step to overcoming it. Take some time to reflect on potential challenges or fears you may face in your transition. This could be fear of failure, the unknown, or needing to improve in a new career.

Consider the importance of setting clear goals and creating a plan to help you and your family navigate through the terror barrier. Setting realistic and achievable goals will give you a sense of direction and purpose during your career transition. Create a step-by-step plan outlining the actions you need to take to reach your goals. This plan can serve as a roadmap to guide you through the challenges that may arise as you move forward. The key is to take action on plans!

Building a support network to help you through your career transition is crucial. Surround yourself with people who believe in you and your abilities. Seek out mentors, career coaches, or supportive friends and family members who can offer guidance, advice, and encouragement when you face the terror barrier. A strong support network can motivate and reassure you to push through your fears and continue moving towards your goals.

Embrace the discomfort that comes with facing the terror barrier. Understand that feeling afraid or uncertain is a natural part of the process. Instead of shying away from these feelings, acknowledge them and use them as fuel to propel yourself forward. As you push through, remember to celebrate small victories along the way. Recognize and appreciate your progress, no matter how small it may seem. Acknowledging your accomplishments will build confidence and resilience, making overcoming future obstacles in your career transition easier.

Action Two: Once You've Identified Your Potential Challenges Or Fears, It's Time To Develop Strategies For Overcoming The Terror Barrier

Now it's time to tackle those fears head-on. One way to do this is through visualization. Block out some time each day to picture yourself conquering those obstacles confidently. See yourself pushing through the challenges and coming out victorious on the other side. This will help train your mind to see success as a real possibility and give you the mental strength to push through the terror barrier.

Another effective strategy is positive self-talk. Replace those negative thoughts with uplifting affirmations and remind yourself of your worth and capabilities. Tell yourself that you can overcome any challenge that comes your way. By boosting your self-confidence and shifting your mindset, you'll be better equipped to face your fears and take the necessary steps toward your goals.

Be bold and lean on your support system during this time. Surround yourself with friends and family who uplift and encourage you. Seek mentors or peers who have gone through similar situations and can offer guidance and support. A strong support system can help bolster your confidence and give you the reassurance you need to push through the terror barrier.

Remember that it's okay to take small steps. You don't have to tackle everything at once. Break down your goals into manageable tasks and celebrate each small victory. Taking it one step at a time will gradually build the confidence and momentum needed to overcome the terror barrier and achieve your desired outcome. Keep pushing forward, and don't let fear prevent you from reaching your full potential.

Next, practice stepping out of your comfort zone in small ways to build confidence. This could be as simple as starting a conversation with someone new or taking on a new challenge at work. Each small step outside your comfort zone will help build your confidence and prepare you for bigger leaps.

Stepping out of your comfort zone can be daunting, but starting small is the key. By practicing stepping out of your comfort zone in small ways, you can slowly build your confidence and push yourself to take on bigger challenges. For example, try striking up a conversation with someone new at a social gathering or volunteer for a task at work that you've never done before. These small acts of courage will help you expand your comfort zone and prepare you for bigger leaps.

When you take small steps outside your comfort zone, you are proving to yourself that you can handle new and unfamiliar situations. Each small success will boost your self-confidence and allow you to step out of your comfort zone again. This process of pushing yourself in small ways will gradually build up your resilience and adaptability, making it easier for you to handle larger challenges in the future.

You also build a growth mindset by consistently practicing stepping out of your comfort zone. You are showing yourself that you are open to learning and growing and willing to embrace new experiences and opportunities. This will benefit you both personally and professionally, as employers value individuals who are adaptable, open-minded, and willing to take on new challenges.

So, the next time you feel hesitant about stepping out of your comfort zone, remember that each small step you take builds your confidence and prepares you for bigger leaps. Embrace the discomfort and push yourself to try new things, and you'll be amazed at how much you can grow and achieve.

Action Three: Find Accountability Partners Or Mentors To Help You Navigate The Terror Barrier

Navigating the terror barrier of a career transition can be over-whelming, but you don't have to do it alone. Seek accountability partners or knowledgeable mentors who can offer support and guidance as you navigate this challenging time. Having someone in your corner who can provide encouragement and wisdom can make all the difference in staying motivated and focused on your goals. Look for mentors or peers who have successfully navigated their career transitions and can offer valuable insights and advice based on their experiences.

Having an accountability partner or mentor can provide valuable encouragement and motivation during times of uncertainty. They can offer a fresh perspective and help you see the situation differently, giving clarity and guidance when needed. Knowing that you have someone to support you can help alleviate some of the stress and anxiety that comes with making a career transition, allowing you to focus on the steps needed to move forward.

Mentors or peers who have successfully navigated their career transitions can offer valuable insights and advice to help you avoid common pitfalls and make informed decisions. They can share their own experiences, provide practical tips, and offer encouragement when you encounter challenges. Surrounding yourself with people who have been through similar experiences and become successful on the other side can give you the confidence and reassurance you need to navigate the terror barrier and become stronger and more resilient.

Accountability partners, mentors, trained coaches can also help hold you accountable for your actions and goals. They can help you set realistic, achievable milestones and keep you on track as you work towards your objectives. Having someone to share your progress with and who can hold you accountable can help you stay focused and committed to your career transition, increasing your chances of success.

In addition to the information provided here, some resources and tools can further help you harness Intentionality and push through the terror barrier. You can find books or articles on the concept of the terror barrier and how it relates to career transitions. There are also visualization apps or techniques that can aid in overcoming the terror barrier. You may also benefit from confidence-building exercises and self-help resources to assist you in stepping out of your comfort zone. Furthermore, online platforms exist for finding accountability partners or mentors who can help navigate the terror barrier. Some examples include:

- Books or articles on the concept of the terror barrier and its relation to career transitions: Please trust me. Look for literature that delves into the psychology behind the terror barrier and offers practical advice on pushing through it in the context of making a career change.

- Visualization apps or techniques: Visualizing success and envisioning yourself on the other side of the terror barrier can be a powerful tool for building confidence and motivation. Look for apps or techniques that help guide you through this process.

- Self-help resources: Countless self-help books, podcasts, and online courses offer advice and strategies for overcoming personal obstacles and barriers. Look for resources that specifically address the concept of the terror barrier.

- Online platforms for finding accountability partners: Connecting with others working to overcome the terror barrier can provide support and accountability. Look for online communities or platforms where you can connect with others facing similar challenges.

- Therapy or counseling: If you're struggling to overcome the terror barrier, seeking professional help can be incredibly beneficial. Therapists and counselors can offer personalized support and strategies for navigating personal obstacles.

Dr. Thurman Fleet, a chiropractor and natural health advocate, invented the Stickman during World War II to help military members maintain their health and wellness. The Stickman, a simple, stick-figure representation of a human body, was used to educate soldiers on the importance of proper posture, alignment, and spinal health. Dr. Fleet understood the physical demands and challenges faced by military personnel, and he believed that by teaching them how to care for their bodies, he could help improve their overall well-being and performance on the battlefield.

In order to change our results and achieve what we want, we must also be willing to change our habits. Whether it's improving our health, excelling in our careers, or cultivating better relationships, our habits play a crucial role in shaping our outcomes. We need to be mindful of our daily routines, behaviors, and thought patterns and be willing to adjust to achieve our desired goals. This may require breaking free from old, limiting habits and adopting new, empowering ones that serve our best interests. It takes going through the terror barrier.

Ultimately, Dr. Thurman Fleet's invention of the Stickman reminds us that small, simple tools and practices can make a big difference in our lives. By adjusting our habits, beliefs, and behaviors, we can transform our outcomes and create the lives we truly desire. By understanding and harnessing intentionality to go through the terror barrier with an understanding of the "stickman" concept, you will be better prepared to tackle the challenges of your military transition and move confidently into a new career or entrepreneurship. Remember, you have what it takes to overcome the terror barrier and become stronger on the other side.

Chapter Summary:

- Understand the terror barrier and how it may appear in your life.

- Develop strategies for overcoming the terror barrier, such as visualization and positive self-talk.

- Practice stepping out of your comfort zone in small ways to build confidence.

- Seek out accountability partners or mentors for support and guidance.

- Utilize resources and tools such as books, apps, and online platforms to assist in overcoming the terror barrier.

Chapter FAQs:

What is the concept of the terror barrier, and how does it relate to career transition?

The terror barrier is the feeling of fear and uncertainty that holds us back from making a change or taking a big step. That voice in our head says, "You can't do it," or "What if you fail?" In career transition, the terror barrier might manifest as fear of failure, the unknown, or not being good enough in a new career.

How can I identify potential challenges or fears related to my career transition?

Take time to reflect on your potential challenges or fears in your transition. This could include taking an inventory of your current worries and concerns, discussing with a career coach or mentor, or conducting self-assessment exercises to identify potential obstacles.

What are some effective strategies for overcoming the terror barrier?

One effective strategy is visualization. Take some time each day to visualize yourself succeeding in your transition. Another strategy is positive self-talk. Replace negative thoughts with positive affirmations and remind yourself of your strengths and capabilities.

What are some examples of stepping out of my comfort zone in small ways to build confidence?

This could be as simple as starting a conversation with someone new, taking on a new challenge at work, or attending a networking event. Each small step outside your comfort zone will help build your confidence and prepare you for bigger leaps.

How can I find accountability partners, mentors, or trained coaches to help me navigate the terror barrier?

Look for mentors or peers who have successfully navigated their career transitions and can offer valuable insights and advice. You can also seek accountability partners through networking events, professional organizations, or online platforms facilitating mentorship connections.

What resources and tools can help me harness Intentionality and push through the terror barrier?

There are books or articles on the concept of the terror barrier and how it relates to career transitions. There are also visualization apps or techniques that can aid in overcoming the terror barrier. Confidence-building exercises and self-help resources can also assist you in stepping out of your comfort zone.

How do I find online platforms for finding accountability partners or mentors?

Websites like LinkedIn, Meetup, or industry-specific forums and groups can be great places to connect with potential mentors or accountability partners who can help you navigate the terror barrier.

What are some examples of confidence-building exercises that can help me step out of my comfort zone?

Role-playing scenarios, public speaking practice, and seeking out constructive feedback are all examples of confidence-building exercises that can help you build the courage to step out of your comfort zone.

Chapter Six

You are Worth it!

STEP-FIVE
Create Your Roadmap to
Personal and Professional Success

This Chapter Will Help You:

- Reflect on your talents, passions, and values to identify your new life purpose.

- Set specific career or business goals aligned with your values and beliefs.

- Pursue education or training opportunities to develop your skills or learn something new.

- Network and seek out opportunities that align with your new purpose.

- Celebrate milestones and successes in your career transition journey.

"To be successful at anything, you don't have to be special. You simply have to be what most people are not. Consistent, determined, and willing to work for it."
~ Jim Rohn

From Successful Lawyer to Deployed Airman: The Shocking Story of Sacrifice and Redemption

David was a low-ranking enlisted member of the Air Force Reserve who had always led a "double life." He was the proud owner of a successful law firm and served as a plumber in the Air Force Reserve component. It was an unconventional situation, but it worked for him. A few years after enlisting, his life turned, and he had to make tough decisions.

David's life was turned upside down when he received orders for a deployment to the Middle East. The news hit him like a freight train, leaving him feeling shocked. Not only was he facing the prospect of being away from home for 24 long months, but the financial implications of his deployment were crippling. As the family's sole breadwinner, David's deployment meant a loss of 90% of his income. He was facing the thought of uprooting his family and jeopardizing his civilian career.

As the reality of the situation sunk in, David grappled with the weight of his responsibilities. Not only was he facing a devastating loss of income, but his office staff, who had become like family to him, were also at risk of losing their jobs. To make matters worse, his wife had to put her education on hold to support the family in his absence. David felt the rug had been pulled out from under him, and he struggled to see a way out of the dark cloud that loomed over his family.

Despite the overwhelming odds stacked against him, David refused to let despair take over. As he prepared for deployment, he arranged for his law firm to continue operating in his absence. He reached out to a couple of trusted friends willing to step in and take care of his clients, ensuring that his staff would not lose their jobs. It was a great victory amid a chaotic situation; it gave David hope that they could weather the storm.

When the day finally came for David to leave for his deployment, his heart was heavy with excitement to serve his country, but still worried for his family and his business. As he settled into his duties overseas, something remarkable began to happen. Despite the hardships and the sacrifices, he was making, David found joy! He had a sense of purpose and fulfillment as a low-ranking enlisted member. He threw himself into his work with unwavering dedication and discovered a profound pride in being part of something bigger than himself.

As the months passed, David's deployment proved a great success. He embraced the notion of service before himself, and his peers, senior enlisted, and officers took notice of his unwavering commitment. They marveled at his willingness to sacrifice his civilian career and finances for the greater good, and David became a shining example of selflessness in the face of adversity.

When the time finally came for David to return home, he was a fulfilled man. The experience had opened his eyes to the true meaning of purpose and satisfaction. He realized that money, title, position, and rank were not the be-all and end-all of a fulfilling life. With an incredible sense of clarity, David retired after serving for 23 years; two years later, he sold his law firm.

Now, David is a coach and trainer, dedicating his time to helping transition military members find confidence as they move into new careers or entrepreneurship. He has found his calling in guiding others through the challenges of transition, and he looks back on his journey with gratitude. Though the road had been fraught with obstacles, David emerged from the experience with a renewed sense of purpose and a deep appreciation for the true meaning of service and sacrifice.

As We Move Forward in This Chapter, These Are Some Of My Recommendations:

Strengths assessment tools: These tools are critical for helping transitioning military members reflect on their talents and passions and essential for identifying their new life purpose and career goals.

Online courses or training programs: Pursuing education and further developing skills is crucial for transitioning military members as they work towards achieving their specific career or business goals aligned with their values and beliefs.

Professional networking platforms: Networking is vital to success in the transition and seeking opportunities aligned with their new purpose. Professional networking platforms help transitioning military members connect with others in their desired industry and gain valuable insights, mentorship, and opportunities.

Personal development books or resources: Celebrating milestones and successes in the career transition journey is essential for maintaining motivation and acknowledging progress. Personal development books or resources can inspire and guide you in navigating this journey.

By now, you've learned a lot about transitioning from the military. Now, it's time to think about what success looks like for you and how to achieve it. In this chapter, we will dive into how to find success and a new life purpose as you transition into civilian life.

Personal Note: When transitioning from the military, one of the most important things to remember is to have confidence in your expectations. I know it is a challenging and daunting time, but having a positive and confident outlook on what lies ahead can make all the difference. Trust in your abilities, skills, and experiences gained from the military, and believe you can succeed in your civilian life. Embrace the unknown with a sense of optimism and determination, knowing that you have the strength and resilience to navigate through any obstacles that may come your way. What helped me in my transition was allowing myself to unlearn and learn something new. After all, the military instilled in us certain habits, routines, and ways of thinking that may need to be adjusted as we enter the civilian world. Be open to new perspectives, ideas, and ways of doing things. Please recognize that it's okay to let go of old habits that no longer might serve you and be open to the growth that comes with learning something new.

Action One: Reflect On Your Talents, Passions, And Values To Identify Your New Life Purpose

One of the first steps in finding success in your transition is to reflect on what makes you unique. What are you good at? What brings you joy? What do you value most in life? Understanding these aspects of yourself, you can identify your new life purpose. For example, suppose you're passionate about helping others and have a talent for problem-solving. In that case, you might discover that a social work or counseling career aligns perfectly with your values and passions.

When reflecting on your unique qualities, it's essential to consider what makes you stand out. You may have a knack for creativity, a love for nature, or a talent for connecting with people. These are all important clues that can help guide you in your transition. By recognizing and celebrating your strengths, you can see the potential for new opportunities and paths that align with your passions.

In addition to recognizing your strengths, it's essential to consider what brings you joy and fulfillment. What activities or experiences make you feel the most alive? What things light up your spirit and make you feel like your best self? By understanding what brings you joy, you can envision your ideal life and start taking steps to make that vision a reality.

Another important aspect of reflection is identifying your core values. What principles and beliefs are most important to you? Maybe you value honesty, empathy, or creativity. Whatever they may be, recognizing and honoring these values can help guide you toward a new purpose that feels meaningful and fulfilling. For example, if you highly value social justice, consider pursuing a career in advocacy or non-profit work.

As you reflect on your unique qualities, passions, and values, remember that this process is about embracing your true self and envisioning a life that aligns with who you are at your core. By reflecting, you can gain clarity and confidence in pursuing a new purpose that brings you a sense of fulfillment and joy.

Action Two: Set Specific Career Or Business Goals Aligned With Your Values And Beliefs

Once you've identified your new life purpose, setting specific career or business goals that align with your values and beliefs is essential. These goals will help guide your transition and keep you focused on what you want to achieve. For example, if you've discovered your passion for technology and innovation, your goal might be to start your own technology company focused on creating solutions for the healthcare industry.

Once you've identified your new life purpose, setting specific career or business goals that align with your values and beliefs is essential. These goals will provide you with a roadmap for your journey and keep you on track as you work towards achieving your dreams. For instance, if you have determined that you are passionate about environmental sustainability, your goal might be to pursue a career in renewable energy or start a business specializing in eco-friendly products and services. By outlining your ambitions in detail, you can gain clarity and direction in pursuing a meaningful and fulfilling career path.

Setting these career or business goals is crucial to ensure they are realistic, achievable, and aligned with your skills and resources. It's important to consider any necessary steps or milestones to be accomplished to reach your ultimate goal. For example, suppose your goal is to become a successful entrepreneur in the fashion industry. In that case, you should focus on gaining relevant experience, building a professional network, and developing a solid business plan. Setting clear, achievable goals will motivate you to act and move forward with purpose and determination.

Additionally, your career or business goals should be aligned with your values and beliefs and contribute to positively impacting the world. Integrating your passion and strengths into your career or business goals will not only bring you a sense of fulfillment and satisfaction, but it will also inspire others and create a ripple effect of positive change. This could mean setting goals focusing on sustainability, equality, social impact, or innovation, depending on what matters most to you and your vision for the future.

Ultimately, setting specific career or business goals that align with your life purpose will help you stay focused, motivated, and driven as you build the life you truly desire. Regularly reviewing and adjusting your goals and staying open to new opportunities and possibilities can create a clear path for yourself and take meaningful steps toward realizing your dreams. You can turn your passion and purpose into a successful and fulfilling career or business venture with the right mindset and determination.

Action Three: Pursue Education Or Training Opportunities To Develop Your Skills Or Learn Something New

Pursuing education or training opportunities to develop your skills or learn something new may be necessary to succeed in your new career or business venture. This might mean enrolling in online courses or training programs to build your expertise further in a particular area. For instance, if you're interested in starting a business in the fitness industry, you should pursue a certification in personal training or sports nutrition to build your expertise and credibility.

If you need more time to get started, consider contacting mentors or professionals in your field for guidance. Networking with others who have achieved success can provide valuable insights and advice as you navigate your new career or business venture. Be bold and ask questions or seek opportunities to learn from those who have already paved the way. Building a solid support network and seeking mentorship can help you feel more confident and prepared as you embark on this new journey.

In addition to seeking educational opportunities and mentorship, staying up to date with industry trends and developments is essential. Technology and best practices constantly evolve, and staying informed is crucial to remain competitive. This might mean attending industry conferences, subscribing to relevant publications, or following thought leaders in your industry on social media to stay informed about the latest advancements and trends. Continuously educating yourself and staying in the know will help you stand out and make informed decisions as you build your career or business.

As you pursue further education and seek out mentorship and industry knowledge, setting clear goals and benchmarks for your progress is essential. Take the time to outline what you hope to achieve in your new career or business venture and break down those goals into manageable steps. This can help you stay focused and motivated as you work towards achieving your aspirations. By setting measurable goals and tracking your progress, you can feel a sense of accomplishment as you progress in your career or business.

Ultimately, the key to succeeding in any new career or business venture is to remain open to new opportunities and experiences. Be willing to adapt and grow as you continue to learn and develop your skills. Embracing a continuous improvement mindset and a willingness to seek new knowledge will help you thrive in your new endeavor. Remember that success doesn't happen overnight, but with dedication, perseverance, and a commitment to continued learning, you can achieve your goals and find fulfillment in your new career or business venture.

Action Four: Network And Seek Out Opportunities That Align With Your New Purpose

Networking is an essential part of finding success in your transition. By connecting with others in your desired industry, you can gain valuable insights, mentorship, and opportunities that align with your new purpose. Whether attending industry-related events or joining professional networking platforms like LinkedIn, actively networking can open doors to new career or business opportunities.

Networking is about more than just exchanging business cards and connecting with people on LinkedIn. It's also about building meaningful relationships with individuals who can support and guide you on your new journey. When you attend industry-related events or join professional networking platforms, go to form genuine connections, and learn from others. Listen to their stories, ask questions, and share your goals and aspirations. Remember, networking is a two-way street, so be prepared to offer support and assistance to others as well.

Mentorship is another crucial aspect of networking during a transition. Finding someone succeeding in your desired industry can provide invaluable guidance and insights. A mentor can offer advice, share their experiences, and help you navigate the challenges and opportunities that come with your new purpose. Seek out mentors who inspire you and whose values align with your own. Building a strong mentor-mentee relationship can significantly enhance your journey and contribute to your overall success.

In addition to gaining insights and mentorship, networking can lead to exciting career or business opportunities. By getting to know people in your industry, you may come across job openings, collaborative projects, or potential clients that align with your new purpose. Stay open to new opportunities and be proactive in pursuing them. Whether through informational interviews, workshops, or industry-related discussions, actively engaging with your network can lead to exciting prospects that move you closer to your goals. Remember, networking is about who you know and how you maintain and grow those connections. Keep nurturing your relationships; they may lead you to the next ample opportunity.

Action Five: Celebrate Milestones And Successes In Your Career Transition And Life Journey

As you progress in your career transition journey, it's important to celebrate the milestones and successes along the way. Take the time to acknowledge your hard work and the progress you've made. This could mean treating yourself to a nice dinner after landing your first client for your new business or taking a weekend trip to celebrate completing a significant milestone in your career transition journey.

Celebrating your milestones and successes is essential to your career transition journey. It's easy to get caught up in the day-to-day grind and forget to take a moment to acknowledge how far you've come. Whether it's landing your first client for your new business or completing a major project, taking the time to celebrate is essential for maintaining motivation and keeping a positive mindset. So, go ahead and treat yourself to something special – you've earned it!

It's about more than just the significant achievements. It's essential to recognize the small victories along the way. Maybe you conquered a fear of public speaking or finally perfected a new skill that's been challenging for you. These smaller milestones are just as important as the larger ones and deserve to be celebrated. Find joy in the journey and take pride in every step forward.

Enjoying your successes can also serve to stay connected with your support network. Share your achievements with friends, family, or mentors who have been rooting for you all along. Their encouragement and positive reinforcement can help boost your confidence and propel you forward in your career transition. Celebrating with others can make the experience even more meaningful and enjoyable.

Lastly, make celebration a regular part of your career transition journey. Set small goals and create opportunities to acknowledge your progress along the way. This can help you stay motivated and focused on the goal. Remember, your journey is just as important as the destination, so take the time to celebrate and savor all the victories, big and small.

Suggested Tools And Resources To Help You Get The Result:

- Strengths assessment tools to reflect on your talents and passions.

- Online courses or training programs to give you further skills development.

- Professional networking platforms: Join platforms like LinkedIn to connect with professionals in your desired industry and seek out career or business opportunities.

- Personal development books or resources: Consider reading books or listening to podcasts celebrating milestones and successes in your career transition journey, such as "The Success Principles" by Jack Canfield or "Grit: The Power of Passion and Perseverance" by Angela Duckworth

Following these steps and utilizing the suggested tools and resources, you will be well on your way to finding success and a new life purpose in transitioning from the military into a new career or entrepreneurship. Remember, you are worth it, and your future is waiting for you to seize it!

Chapter Summary:

- Reflect on talents, passions, and values to identify new life purpose.

- Set specific career or business goals aligned with values and beliefs.

- Pursue education or training opportunities to develop skills or learn something new.

- Network and seek out opportunities that align with new purpose.

- Celebrate milestones and successes in your career transition journey.

Chapter FAQs:

How can I reflect on my talents, passions, and values to identify my new life purpose?

Reflecting on your abilities, passions, and values involves taking the time to introspect and consider what brings you joy, what you're good at, and what you value most in life. This could be done through activities such as journaling, meditation, or seeking the input of trusted friends and mentors.

What are some examples of specific career or business goals aligned with my values and beliefs?

Examples of career or business goals aligned with your values and beliefs could include starting a nonprofit organization focused on a cause you deeply care about, pursuing a career in a field that aligns with your values, or starting a business promoting sustainability and environmental conservation.

How can I pursue education or training opportunities to develop my skills or learn something new?

Education or training opportunities may involve enrolling in online courses, workshops, or certification programs related to your area of interest. For instance, if you're interested in a career in digital marketing, you could pursue a certification in Google Analytics or social media marketing.

What are some effective networking strategies to seek opportunities aligning with my new purpose?

Effective networking strategies include attending industry-related events, joining relevant professional associations, contacting professionals for informational interviews, and actively engaging on professional networking platforms like LinkedIn to connect with individuals in your desired industry.

How do I celebrate milestones and successes in my career transition journey?

Celebrating milestones and successes in your career transition journey could involve rewarding yourself in meaningful ways, such as treating yourself to a small indulgence, taking a short break to relax, or celebrating with friends and family to acknowledge your progress.

What are some recommended strengths assessment tools to reflect on my talents and passions?

Strength assessment tools, such as StrengthsFinder, can help you gain insight into your unique strengths and talents, providing valuable information as you identify your new life purpose.

How can I effectively use professional networking platforms like LinkedIn to connect with professionals in my desired industry?

Utilizing professional networking platforms like LinkedIn involves creating a professional and complete profile, actively engaging with relevant content, joining industry-specific groups, and reaching out to individuals for informational interviews or mentorship opportunities.

What are some ways to stay motivated and focused on my journey to finding success and a new life purpose?

Staying motivated and focused may involve setting small, achievable goals, seeking support from mentors or a peer group, maintaining a positive mindset, and regularly reminding yourself of your long-term vision and purpose.

Conclusion

Congratulations on finishing "Take Charge of Your Military Transition." I hope you found the book to be a valuable resource as you navigate this significant life change.

As a 32-year veteran who has experienced the transition from military to civilian life firsthand, I understand the challenges and uncertainties of this process. That's why I wrote this book - to provide you with the mentorship and guidance I wish I had when I was in your shoes. My goal is to help you avoid wasting precious time, effort, and energy by placing them in the wrong place or with the wrong people. I've shared practical advice, tips, tricks, and examples to help you master this important topic and set yourself up for success in the civilian world.

Now that you've reached the end of the book, it's time to put what you've learned into action. Transitioning from the structured environment of the military to the fast-paced and ever-changing civilian world can be daunting, but with the right mindset and preparation, you can thrive in this new chapter of your life.

The first step is to act. It's easy to feel overwhelmed by the prospect of starting over, but you'll gain momentum and confidence by taking proactive steps towards your goals. Whether it's updating your resume, reaching out to potential employers, or networking with other veterans who have successfully trans-itioned, every small action can lead to big results.

Next, take the time to assess your mindset and emotions. Transitioning from the military can bring up a range of emo-tions, from excitement and anticipation to fear and uncertainty. Acknowledging and processing these feelings to approach your transition with a clear and focused mindset is important.

Understanding your personality under stress is also crucial. The skills and traits that served you well in the military may need to be adapted for civilian life. By recognizing your strengths and areas for growth, you can tailor your approach to the job market and position yourself for success.

Finally, make an effort to network in civilian settings. The connections you make during your transition can be invaluable as you seek out new opportunities. Attend networking events, join professional organizations, and reach out to individuals in your desired industry. You never know where your next opportunity may come from, so casting a wide net and building a strong professional network is key.

Transitioning from the military to civilian life is no small feat, but it can be a rewarding and fulfilling experience with the right mindset and approach. Remember, you have the skills, experi-ence, and determination to succeed in any setting. Putting the advice and guidance from "Take Charge of Your Military Transition" into action, you'll be well on your way to finding incredible civilian opportunities and charting a successful path forward.

So, what are you waiting for? It's time to act, assess your mindset and emotions, understand your personality under stress, and start networking in civilian settings. The civilian world is waiting for you, and I have no doubt that you have what it takes to conquer this new challenge and thrive.

Best of luck in your transition and thank you for your service. The next chapter of your life is full of potential - embrace it and make it your own! Do not hesitate to contact me if you have any questions or need anything from me.

Ericka E. Kelly, Ret CMSgt

As a 32-year veteran of the military, Ericka E. Kelly has seen and experienced it all when it comes to transitioning from a military career to civilian life. Her latest book, "Take Charge of Your Military Transition," is a guide for transitioning military members who are seeking clarity and guidance during this pivotal time in their lives.

Ericka E. Kelly wrote this book to provide a different roadmap for military members who want to transition well from their military careers and find incredible civilian opportunities. With her extensive experience and knowledge, she aims to empower her readers to navigate this transition with confidence and determination.

She is uniquely qualified to help transitioning military members reach their goals. With her background as the Senior Enlisted Advisor to the Chief of the Air Force Reserve and Command Chief Master Sergeant for the Air Force Reserve Command, she offers valuable insights and practical advice.

Additionally, as a European Mentoring and Coaching Council (EMCC) Master Coach, she is an iconic mindset expert. With her Certificates as a John C. Maxwell speaker, trainer, and licensed DISC trainer and consultant, she brings a wealth of expertise in communication skills and personality styles.

Her civilian professional background includes serving as a Senior Special Agent for the Departments of Justice and Homeland Security. Notably, Ericka was honored by the United States Congress for her invaluable contributions to the nation's security.

Ericka E. Kelly's educational achievements include a bachelor's degree in criminal justice and a master's in business administration.

If you're a transitioning military member looking for clarity and guidance, Ericka E. Kelly and this book are the perfect resources to help you on your journey. It's time to take charge of your military transition and uncover the incredible civilian opportunities that await you.

"Take Charge of Your Military Transition" was created for you!

Glossary

Adaptability — The ability to adjust to new circumstances and change. This is crucial for transitioning military members as they adjust to civilian life and potential career changes.

Assimilation — The process of adapting or integrating into a new environment. Need to assimilate into civilian work culture and society as they transition out of the military.

Beliefs — Personal convictions and principles that guide behavior and decision-making. Need to reevaluate beliefs and adapt them to fit a new civilian career path.

Career Counseling — Guidance and support provided to individuals in making career-related decisions and transitions. This offers transitioning military members access to expert advice and resources for navigating their civilian career paths.

Career Transition — The process of moving from a military career to a civilian career. This is the primary focus for transitioning military members seeking new opportunities.

Civilian Opportunities — Potential career paths and job openings available to transitioning military members in the civilian sector. Understanding these opportunities is essential for a successful transition.

Civilian Workforce — The non-military workforce of a country or region. Transitioning military members must understand the culture, dynamics, and opportunities within the civilian workforce.

Commitment — A pledge or dedication to a cause or goal. Need to demonstrate commitment to transition and post-military success.

Communication Skills - The ability to effectively convey information and ideas. Strong communication skills are important for networking, job interviews, and workplace success in civilian careers.

Education Opportunities — Programs and resources for furthering knowledge and skills through academic or vocational training. This is important for transitioning military members to explore and pursue educational opportunities that can enhance their civilian career prospects.

Flexibility — Adapting to different situations and changing course as needed. Transitioning military members must be flexible as they explore new career paths and opportunities.

Job Search Strategies — Planned approaches and methods for seeking and securing employment. This helps transitioning military members effectively navigate the civilian job market and secure meaningful employment.

Limitations — Restrictions or obstacles that hinder progress. May need to identify and address any limitations that could impact their transition and pursuit of civilian opportunities.

Mentorship — Guidance and support provided by an experienced and trusted advisor. Transitioning military members can benefit from mentorship to gain insight and advice from someone with civilian transition experience.

Mindset — A person's attitude and way of thinking. Need to develop a positive and adaptable mindset to navigate the transition process and embrace new opportunities effectively.

Networking — The act of building relationships and connections within a professional community. Effective networking is crucial for finding civilian opportunities and mentorship.

Paradigm — A model or pattern of something. May need to shift their paradigm and way of thinking to transition and succeed in a civilian career effectively.

Personality — The combination of characteristics and qualities that form an individual's distinctive character. Need to understand personality traits and how to leverage civilian opportunities.

Professional Development — Activities and resources to enhance professional skills, knowledge, and effectiveness. Essential for transitioning military members to improve and adapt to the civilian workplace continuously.

Resources — Tools, support, and assets that can be used to achieve a goal. May need to access various resources, including mentorship and guidance, to transition and pursue civilian opportunities successfully.

Self-motivation — The drive to pursue goals and work independently. Self-motivation is important for transitioning military members as they navigate their career transition and pursue new opportunities.

Skill Transferability — The ability to apply skills and experiences from military service to civilian careers. Understanding skill transferability is important for identifying potential career paths.

Success — Achievement of desired goals or outcomes. Seek to succeed in a civilian career by effectively transitioning from the military and seizing new opportunities.

Time Management — The ability to prioritize tasks and manage one's time effectively. Strong time management skills are important for transitioning military members as they navigate their career transition process.

Transferable Skills — Skills and abilities acquired during military service that can be applied to civilian jobs. This helps transitioning military members identify and articulate their valuable skills to potential employers.

Transition — The process of moving from one phase or stage to another, such as leaving the military and entering civilian life. How to properly navigate the changes and challenges of transitioning from a military career to civilian opportunities.

Values — Principles and standards that are important to a person. Need to align values with post-military goals to find satisfaction and success in civilian life.

Veterans' Benefits — Financial, educational, and healthcare resources provided to former military members. This is important for transitioning military members to understand and leverage the benefits available to them as they enter civilian life.

Work-life Balance — The equilibrium between professional and personal responsibilities and activities. A crucial consideration for transitioning military members as they strive to maintain a healthy balance in their new civilian careers.

Additional Resources

Looking to make a smooth and fulfilling shift into civilian life? Look no further! Our Transitioning Coaching Sessions, Group Mentoring, and Facilitated Classes are designed to help you confidently navigate this important life change and succeed.

- Our Self-Growth and Transition Masterclasses provide you with valuable tools and strategies to help you transition well from your military career and discover incredible civilian opportunities.

- Through our DISC Assessments, you will gain insight into your unique personality and behavioral style, allowing you to understand yourself better and communicate effectively with others.

- Complimentary monthly Leadership Military Sessions will help you leverage and apply your military leadership skills to the civilian world.

- Complementary quarterly DISC Explanation Sessions will help you understand different personalities and decide whether to purchase an assessment.

Don't let the transition from military to civilian life overwhelm you. Join our programs today and take control of your future! Unlock your full potential and thrive in your new civilian career.

Contact me now to learn more about our upcoming sessions and take the first step toward a successful transition.

Text me: 562-896-9484

ericka.kelly@ekenterprises.co

https://erickakellyenterprises.com/

Recommended Books

In the world of self-growth and mindset, some books have impacted me the most in my transition journey. Each of these influential books offers valuable insights and strategies for personal growth, urging readers to tap into their full potential and live their best lives.

1. "Habits of Highly Effective People" by Stephen R. Covey — It has stood the test of time and has helped readers not only in their professional lives but also in their personal lives, emphasizing the importance of integrity, honesty, and collaboration.

2. "Atomic Habits" by James Clear — Clear's approach to the science of habits and how small changes can lead to remarkable results has inspired readers to take control of their habits and ultimately, their lives.

3. "The Power of Now" by Eckhart Tolle — Is a profound exploration of mindfulness and being present in the moment, a powerful reminder to live in the now rather than dwell on the past or worry about the future.

4. "Think and Grow Rich" by Napoleon Hill — Is a classic in the world of personal development, focusing on the power of positive thinking and the mindset required for success.

5. "Essentialism" Greg McKeown – It emphasizes the importance of prioritizing and focusing on what truly matters.

6. "You Are A Bad Ass" by Jen Sincero — Is a bold and sassy guide to embracing your true potential and living an unapologetically awesome life.

7. "Outliers" by Malcolm Gladwell — delves into the factors contributing to success, challenging conventional wisdom and shedding light on the true nature of achievement.

8. "The Four Agreements" by Don Miguel Ruiz. — provides practical wisdom based on ancient Toltec teachings, offering a powerful code of conduct to guide readers toward personal freedom and happiness.

www.ingramcontent.com/pod-product-compliance
Lightning Source LLC
Chambersburg PA
CBHW060945040426
42445CB00011B/1007